What is a Hot Spring?

BY JACOB SMITH

What is a Hot Spring?

Copyright © 2024 Jacob Smith
All rights reserved.

For more information contact neverendingfieldtrip@gmail.com

ISBN: 9798344183978

From the colorful springs and violent geysers of Yellowstone National Park to the peaceful *onsen* spas of Japan, hot springs have been a source of fascination around the world for thousands of years.

They are found across every continent on Earth and range from small, lukewarm springs to huge pools of boiling hot water.

But what are hot springs, really? Why are they hot? And most importantly, why does it even matter?

Let's take a closer look at each of those questions and find out more about hot springs!

Throughout the book, I will have boxes like this one to explain some terms you may not know.

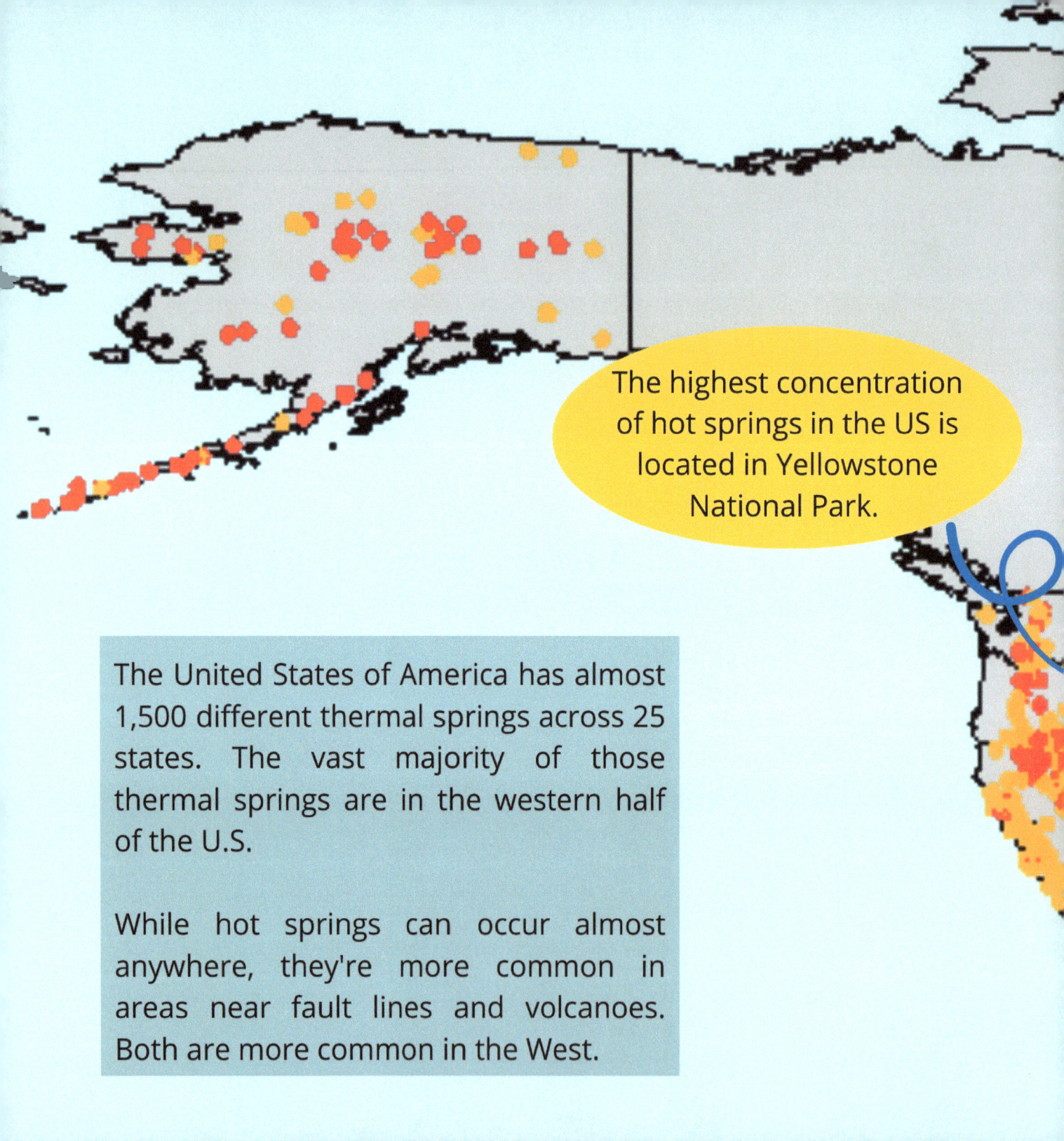

The highest concentration of hot springs in the US is located in Yellowstone National Park.

The United States of America has almost 1,500 different thermal springs across 25 states. The vast majority of those thermal springs are in the western half of the U.S.

While hot springs can occur almost anywhere, they're more common in areas near fault lines and volcanoes. Both are more common in the West.

The springs on this map are divided into "Hot" springs and "Warm" springs.

The orange dots represent the warm springs, which are between 68°F and 122°F.
Red dots represent hot springs, which are above 122°F.

Many of the "hot springs" that people like to visit are really only warm springs. But while these springs aren't as hot as "hot" springs, they are still as hot or hotter than the water in a typical bath. True hot springs have water that is so hot it would burn you!

The water's high temperatures are why we call hot springs "hot" but what about the other part?

What exactly is a **spring**?

Public Domain graphic from
https://commons.wikimedia.org/w/index.php?curid=578349

What is a Spring?

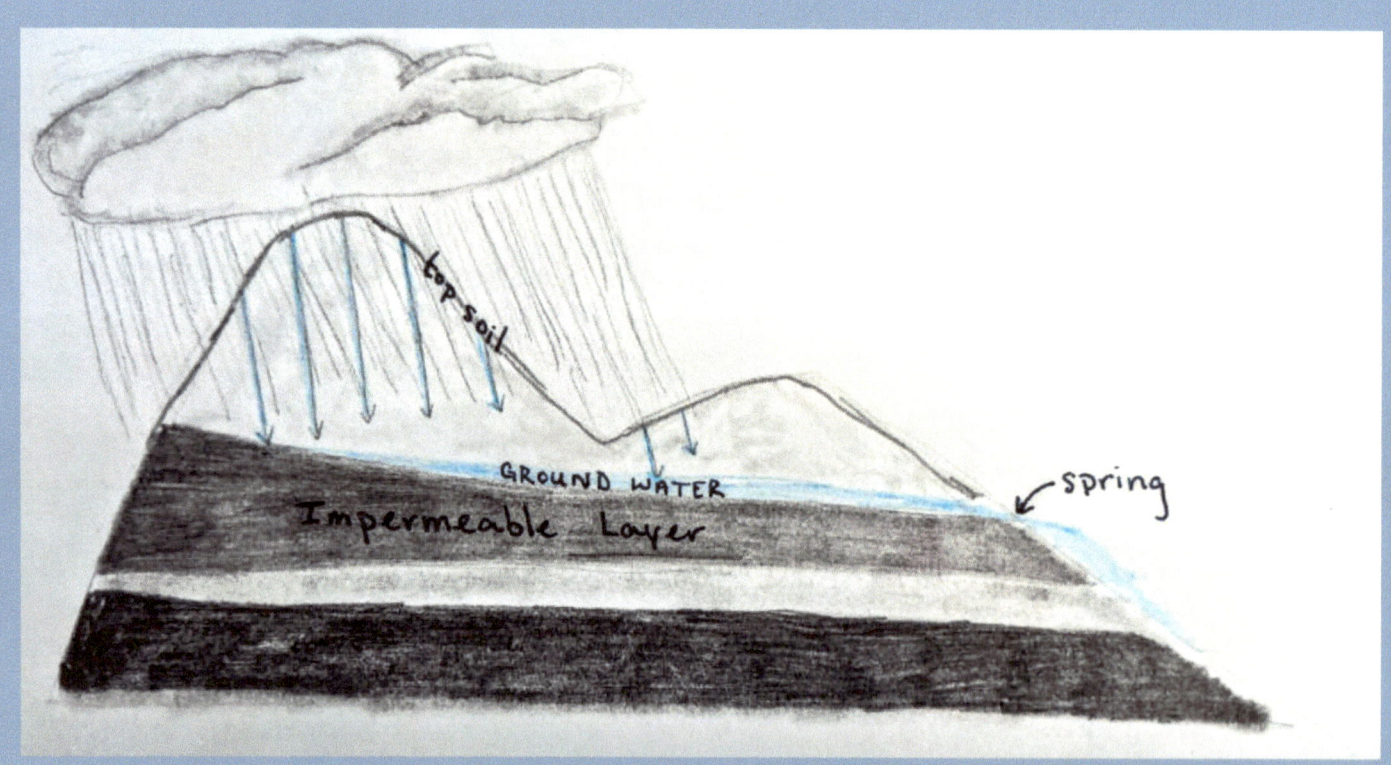

A spring is defined as the place where an underground water source comes to the surface, which sounds simple enough.

But where does the underground water come from in the first place, and how does it end up on the surface?

Springwater typically comes from one of two places. It either falls from the sky as rain, or it comes from melting snow and ice.

In both cases, only a small portion of the water stays above ground as surface water.

The rest soaks into the dirt and becomes groundwater, or the water that we find underground. Once there, the groundwater continues to flow downhill until it either settles in an underground lake, called an aquifer, or until it comes to the surface again.

When it comes back to the surface, we call that flow of water a spring!

Surface water - Water that is above ground, like a river or a lake
Groundwater - Any water that is under the ground
Aquifer - A large body of water under the ground

Gravity Springs

Gravity springs are what most people think about when springs are mentioned. The ground above a gravity spring is composed of sand, soil, or rock with lots of holes in it. These loose layers form a *permeable* layer of ground. Water soaks into this layer until it hits a solid, *impermeable* layer that it can't pass through.

After that, gravity pulls the water downhill until it comes to a place where the permeable layer ends and the impermeable layer is uncovered. The water flows out from the ground, and a gravity spring is formed!

Artesian Springs

Artesian springs are very similar to gravity springs. Rain falls on a permeable layer and sinks into the earth until it meets an impermeable layer just like before. But, instead of escaping, artesian spring water flows underneath a second impermeable layer. When this happens, the groundwater is unable to escape from between the solid layers until it finds a hole in the surface above it.

Interestingly, if the hole in the top layer is smaller than the *recharge area*, the pressure can build up enough to cause the spring to shoot water into the air!

If the hole in the top layer is man-made, it is what is called an artesian well.

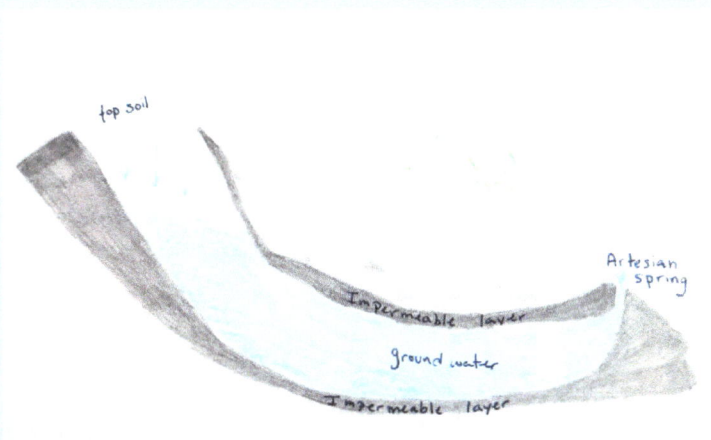

permeable - *allowing substances to pass through easily*
impermeable - *not allowing substances to pass through*
recharge area - *the area where surface water becomes groundwater*

Fracture Springs

These springs occur where a layer of stone that would normally be impermeable has cracks, or *fractures,* where water can pass through easily. Sometimes, those cracks in the rock open onto the surface. If those cracks are below the *water table,* it becomes a fracture spring.

Sometimes, the cracks are actually lava tubes caused by cooled lava flows! When this happens, the fracture spring is instead called a *tubular spring*!

fracture - *a split or separation in a rock layer*

water table - *the line between ground that is filled with groundwater and the dry ground above it*

tubular spring - *a kind of spring where a crack in a hardened lava tube allows water to escape to the surface*

Photo by Sharib4rd | Flickr.com | CC 2.0

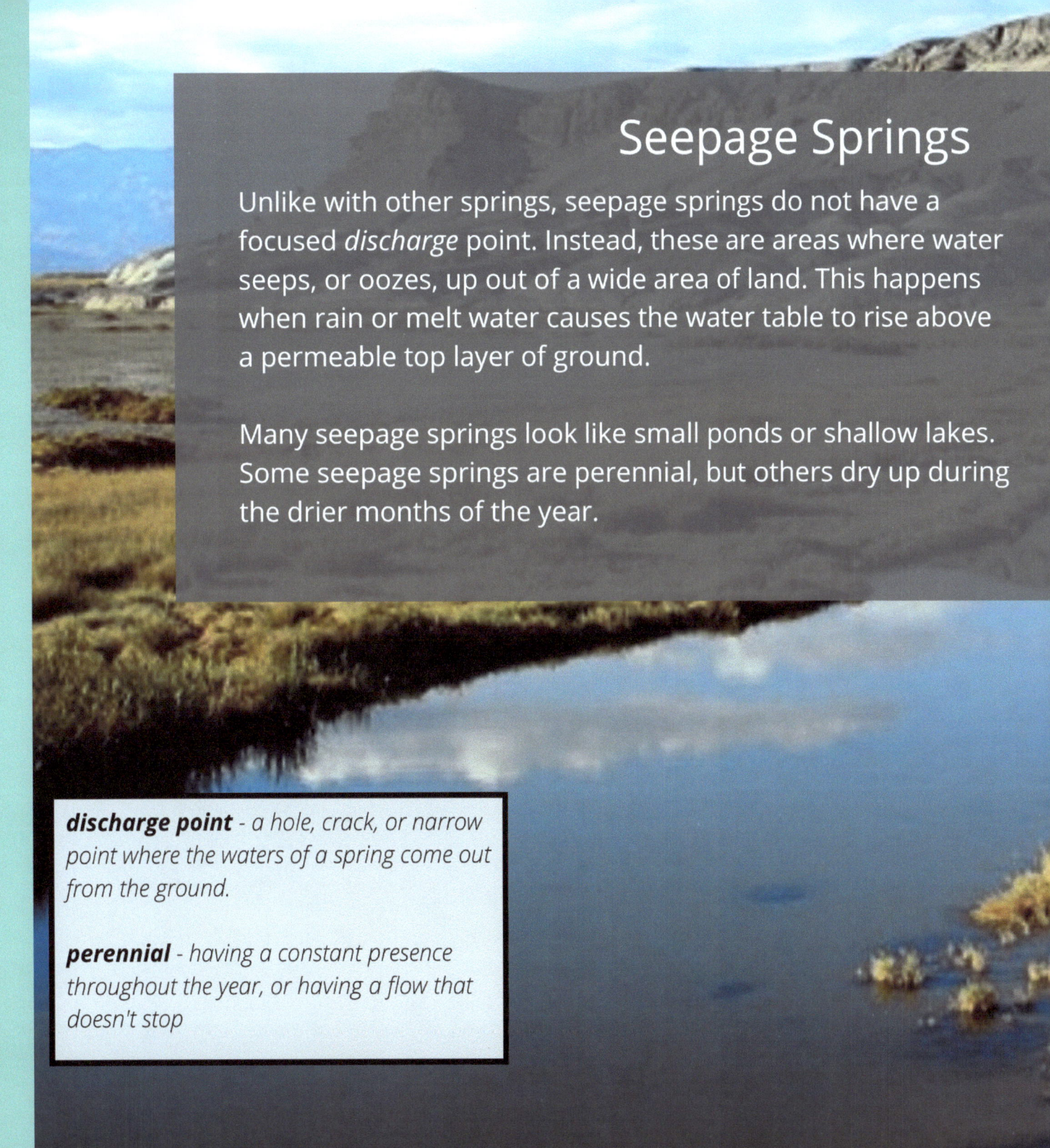

Seepage Springs

Unlike with other springs, seepage springs do not have a focused *discharge* point. Instead, these are areas where water seeps, or oozes, up out of a wide area of land. This happens when rain or melt water causes the water table to rise above a permeable top layer of ground.

Many seepage springs look like small ponds or shallow lakes. Some seepage springs are perennial, but others dry up during the drier months of the year.

discharge point - *a hole, crack, or narrow point where the waters of a spring come out from the ground.*

perennial - *having a constant presence throughout the year, or having a flow that doesn't stop*

Why are hot springs hot?

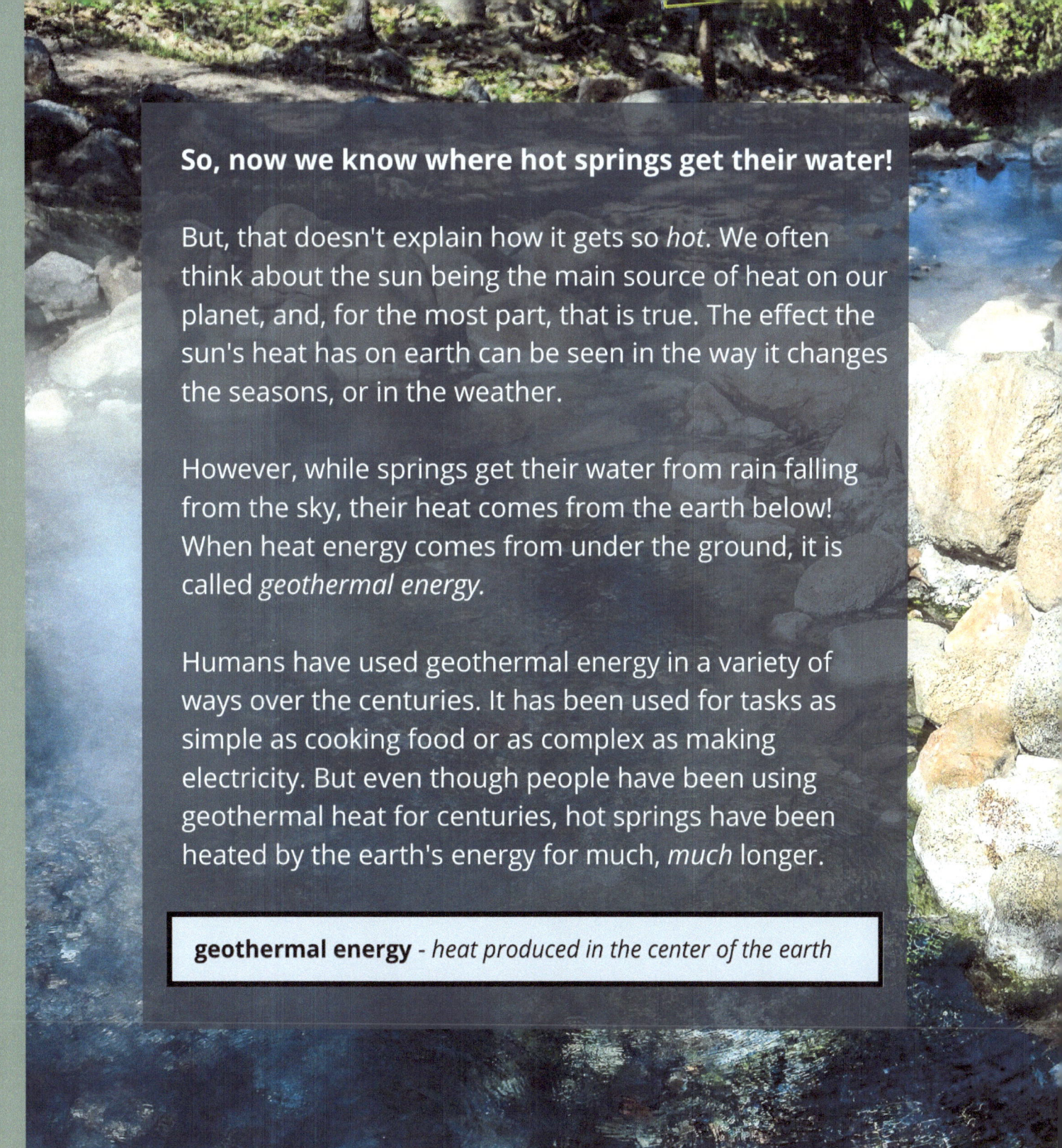

So, now we know where hot springs get their water!

But, that doesn't explain how it gets so *hot*. We often think about the sun being the main source of heat on our planet, and, for the most part, that is true. The effect the sun's heat has on earth can be seen in the way it changes the seasons, or in the weather.

However, while springs get their water from rain falling from the sky, their heat comes from the earth below! When heat energy comes from under the ground, it is called *geothermal energy*.

Humans have used geothermal energy in a variety of ways over the centuries. It has been used for tasks as simple as cooking food or as complex as making electricity. But even though people have been using geothermal heat for centuries, hot springs have been heated by the earth's energy for much, *much* longer.

geothermal energy - *heat produced in the center of the earth*

The outermost layer of the earth is called the crust. While it is significantly thinner than the other layers, it is *easily* deep enough to provide all the geothermal energy needed to create hot springs!

Many hot springs are heated when their waters come in contact melted rock under the ground called "*magma*".

Other hot springs are heated by the ambient heat of the Earth itself.

Magma Chamber

2.5 miles
4km

While we know there are sources of water *extremely* deep under the ground, we don't actually know *how* deep these springs go!

3.7-6.2 miles
6-10km

The earth becomes hot enough to boil water at only around 2.5 miles deep, but many springs are much deeper than that.

Most volcano's magma chambers are found between 3.7 and 6.2 miles underground

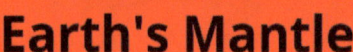

Earth's Mantle

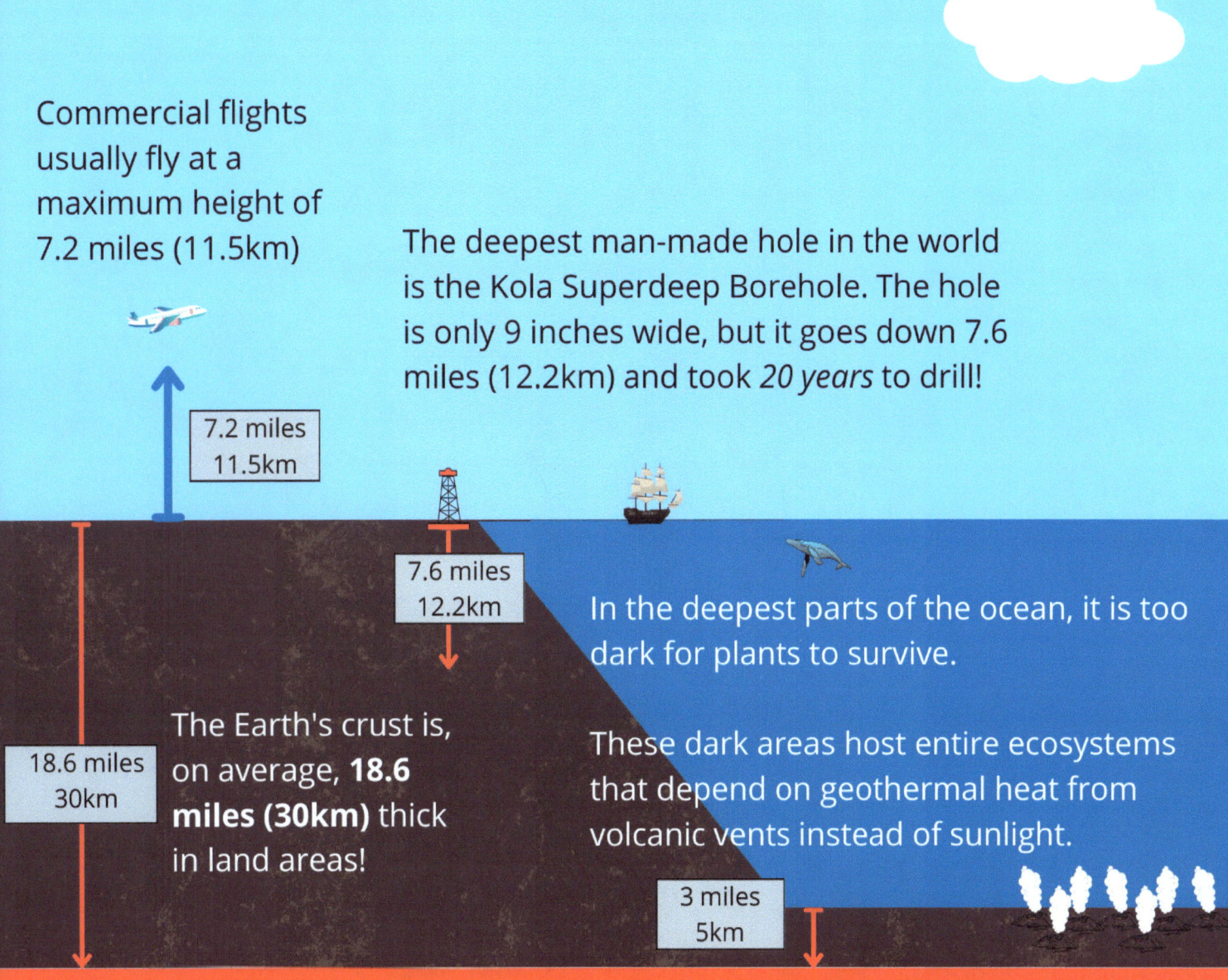

Volcanoes

Volcanoes are one of the most visible signs of the immense heat and pressure found underground.

They form when the hot *magma* from the earth's *mantle* rises through cracks in the crust. Because the magma is under so much pressure, it can come to the surface with some very explosive results!

While there are over 1,000 different volcanoes around the world that could have an *eruption*, there are normally only 50-70 volcanic eruptions that happen each year.

When **magma** makes its way to the surface, it is called **lava!**

mantle - *the layers of the earth underneath the crust*
magma - *hot, melted rock trapped under the ground*
eruption - *an explosive release of gases and other materials, like lava, or hot water*

Geysers

On the surface, geysers and volcanoes can appear similar, but there are some key differences in how geysers work that set them apart from volcanoes.

The biggest difference is that while volcanic eruptions shoot hot rocks and lava into the air, geysers are a special kind of hot spring that shoots hot water into the air.

Geysers are always found in areas with volcanic activity. This is because the heat from magma is needed to vaporize the water quickly enough to erupt!

Unlike volcanoes, it isn't the pressure of the earth's crust forcing the water to the surface. Instead, the pressure from water rapidly becoming steam is what causes a geyser to erupt.

While geysers can only be found near volcanoes, hot springs can be found all over the world, even in places where there aren't any magma pockets at all.

How do these hot springs get their heat?

Hot springs that are far away from magma pockets get their heat from a geothermal effect called the *geothermal gradient*.

Essentially, the deeper you travel under the ground, the hotter it gets. When water travels deep under the ground, it also gets hot. When this hot water comes back to the surface, a hot spring is formed!

geothermal gradient- *the tendency for temperatures to rise as one goes deeper under the ground*

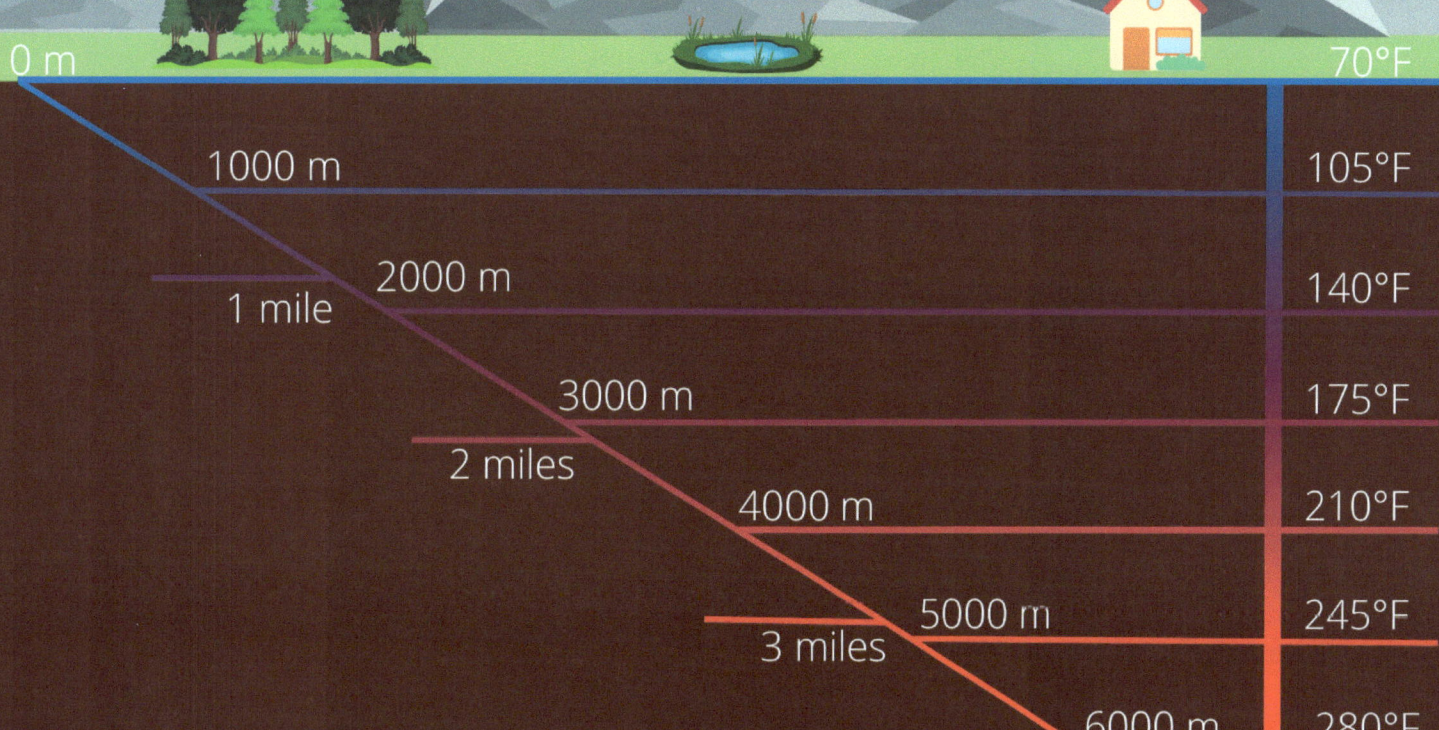

Why are Hot Springs important?

Photo by PunkToad - Flickr: Chena Hot-springs, CC BY 2.0

While they *are* interesting, it may seem like hot springs really aren't all that important. After all, they are basically just large pools of hot water.

But in reality, hot springs are important in a number of different ways.

Tourists relax in a hot spring on Deception Island, located just north of Antarctica

Castle Geyser is a popular tourism spot located in Yellowstone National Park

Tourism

Whether it's looking for a relaxing hot spring to soak in or exploring the geysers of Yellowstone, thermal springs are a major draw for tourism!

Hot springs generate more than **50 billion dollars** in revenue from tourism each year. They are an important economical benefit for the communities where they are found.

Thermal springs attract visitors in a variety of ways, including natural spas, eco-tours, medical tourism, and more.

Essentially, if someone is interested in visiting a thermal spring for a reason other than work, it's tourism!

Geothermal Energy

Another important thing that we get from hot springs is electricity!

While geothermal energy plants produce relatively little electricity compared to older methods, it is a much greener way to make electricity than burning coal or gas.

As technology advances, geothermal energy is becoming *even more* promising as a way to easily and affordably produce electricity.

Someday soon, the electricity you use at home may come from a hot spring!

By Prosthetic Head - Own work, CC BY-SA 4.0

Fast Fact:

Two-thirds of all energy used in Iceland comes from geothermal sources!

Medicinal Uses and Health Benefits

FDR Presidential Library & Museum, CC BY 2.0

President Franklin D. Roosevelt was known for taking swims in thermal springs as a way of helping with his polio.

Over the years, individual hot springs around the world gained a reputation for having healing powers. Spring water was thought to be a cure for all sorts of sicknesses.

Soaking in warm spring waters improves blood circulation, increasing blood oxygen levels. And many people find soaking in a hot spring to be relaxing, lowering the body's stress levels and promoting general good health!

In some places, hot springs water is filled with beneficial minerals from deep inside the earth. Drinking this water can be like taking a vitamin, and it can promote good health! But....

 # Be Careful!

While some hot springs have water filled with beneficial minerals, **not all hot springs are safe to drink**.

Early explorers in Yellowstone Park used the water from the springs to make their coffee only to find out that the spring water there is full of harmful chemicals, including cyanide!

Yikes!

By Chris Light - Own work, CC BY-SA 4.0,

Interested in trying water from a hot spring?

Filling stations, like this one at Hot Springs National Park in Arkansas, allow anyone who wants to fill up a water bottle with fresh, hot water from the springs!

Animals...

Other animals rely on hot springs for their day-to-day survival, too.

The hot spring keelback is a rare species of snake that lives in the Himalayan Mountains of Tibet. It relies on the mountain's hot springs to keep its body warm and would otherwise not be able to survive!

Hot springs aren't just important for humans. Often, the animals that live around hot springs make good use of them, as well.

Walking through Yellowstone National Park, there are many places where you can see buffalo and elk footprints right next to boiling springs.

That's because during the cold winters in the park, these animals use the springs as a way to keep warm!

...and Hot Springs

Probably the most famous example of an animal that uses hot springs is the Japanese macaque, or snow monkey.

These monkeys live high in the mountains of Japan, where snow regularly piles up high. In order to keep warm, they regularly bathe in the hot springs.

There are even **fish** that live in thermal springs!

The Devil's Hole Pupfish lives in a thermal spring in Death Valley, despite water that is often over 90°F, as well as being extremely salty!

Cultural importance of Hot Springs

In many places, thermal springs are also very important culturally.

People across the globe have been visiting thermal springs for thousands of years, and in many places, traditions came up to reflect that.

Japan
Japanese hot springs, or *onsen*, were originally used by Buddhist monks for their purification rituals. Over time, the general population saw the benefit of visiting the springs for their own health and relaxation. Today, they remain an important feature in Japanese culture at large.

Rome
The word "spa" is an acronym for the Latin phrase, **S**anus **P**ar **A**quam, meaning health through water.

Roman bathhouses would often have three separate rooms at different temperatures. Visitors would start with a cold-water bath. They would then move to a warm room to help them adjust to the heat gradually before getting into the hot spring water.

The Americas
Native Americans would often use hot springs for purification rituals, as well as simply to relax. It is thought that they were among the first people to do so, worldwide.

By Noomen9 - CC BY-SA 4.0

About the Author

Hey! I'm Jacob.

I'm a world-schooling dad and the creator of the travel website, the Neverending Field Trip.

Together with my family, I travel the world to find fun, educational experiences with the goal of getting as many families as possible traveling with their kids!

I'm a big fan of hot springs (of course), but I also love old ruins, coral reefs, and exploring everything the world has to offer!

www.ingramcontent.com/pod-product-compliance
Lightning Source LLC
Chambersburg PA
CBHW051822210526
45473CB00005B/1702